The General Herd Book of the Island of Guernsey
Guernsey Cattle – Volume 1

by Thomas Mauger Bichard

with an introduction by Jackson Chambers

Self Reliance Books

Get more historic titles on animal and stock breeding, gardening and old fashioned skills by visiting us at:

http://selfreliancebooks.blogspot.com/

Introduction

I am pleased to present another title in the "Cattle" series.

The work is in the Public Domain and is re-printed here in accordance with Federal Laws.

As with all reprinted books of this age that are intended to perfectly reproduce the original edition, considerable pains and effort had to be undertaken to correct fading and sometimes outright damage to existing proofs of this title. At times, this task is quite monumental, requiring an almost total "rebuilding" of some pages from digital proofs of multiple copies. Despite this, imperfections still sometimes exist in the final proof and may detract from the visual appearance of the text.

I hope you enjoy reading this book as much as I enjoyed making it available to readers again.

Jackson Chambers

COMMITTEE OF HERD BOOK

PRESIDENT:

JOHN BOYD KINNEAR, Esq.

VICE-PRESIDENT:

GEORGE ALLEZ, Esq.

TREASURER:

THOMAS MAUGER BICHARD.

SECRETARIES:

THOMAS MAUGER BICHARD.

HENRI DE MONTEYREMAR.

COMMITTEE:

CHARLES SMITH, Couture, St. Peter-Port.

JOHN OGIER, Devaux, St. Sampson.

THOMAS H. OGIER, Ville Baudu, Vale.

THOMAS LE POIDEVIN, Barras, Vale.

THOMAS LE PELLEY, Beaucamps,
HENRY B. OGIER, Bernauderie, } Castel.
WILLIAM CARRINGTON, King's Mills,

JOHN BROUARD, Choffins,
JOHN A. NAFTEL, Lohiers, } St. Saviour.
PETER LE CHEMINANT, Doult,

MATTHEW TOSTEVIN, Buttes, St. Peter-in-the-Wood.

DENYS CORBET, La Robergerie,
JOHN DE LA RUE, Fontenelles. } Forest.
GEORGE TORODE, Bourgs.

GEORGE ROSE, Saint,
PETER MOLLET, Villette, } St. Martin.

ANTHONY ROBERT, Mauxmarquis,
ROBERT BEST, Brickfield,
BONAMY BROUARD, Courtil-au-Préel, } St. Andrew.
WILLIAM HUBERT, Grand Courtil,

PREFACE.

~~~~~~~

A FEW WORDS are necessary to explain the nature and objects of the Herd Book of which the First Part is now issued, especially in view of the circumstance that another publication bearing the same name, but of quite different character, has already appeared.

The purposes of a Herd Book are, firstly, to register pure bred stock, and, secondly, by recording pedigree to enable the scientific breeder to follow and select the strains of blood which he deems most valuable. In carrying out this design in the leading Herd Books of Great Britain and America it is left to the owners to register such animals as they think fit, subject only to the condition of proving their purity of breed.

But in the Channel Islands, where purity of blood is secured by laws prohibiting importation of breeding stock, it had occurred to some persons to make an entirely different idea form the basis of a Herd Book. Following a suggestion which was first started in Jersey, a few gentlemen in Guernsey, about five years ago, proposed a Herd Book in which the entries were to be limited to animals which they themselves should approve. They inserted some which were already dead, and took for their standard of living ones their own ideas as to external appearance, together with the "milk test" invented by M. Guenon, a Frenchman, who conceived he had discovered an infallible criterion of the quantity of milk a cow would yield in the direction of certain lines of hair between the udder and the root of the tail. Of this discovery it is sufficient to say that whether successful or not with

French breeds it is proved to be entirely fallacious when applied to Guernsey cows. A list of some 170 animals, made up on these principles, was published in 1879 under the name of "The Guernsey Herd Book," and a second part was subsequently issued.

The great majority of Guernsey farmers, however, held aloof from this scheme; and dissatisfaction with the results of the judging caused some of its first supporters to leave it. In 1881 its projectors offered to hand it over to the Royal Agricultural and Horticultural Society of the island. That body, by a narrow majority, in which a number of persons who were not owners of stock were included, decided to take over the publication in so far as it had already proceeded; to add to it its own show prize lists of the two preceding years; and to carry it on for the future by inserting only the names of such animals as might take prizes at future shows. These exceed one hundred in each year, prizes being given to considerably more than half of the number exhibited. But still further to augment the number to be inscribed in the Herd Book as winners of prizes, it was resolved to hold parish shows, at which no money prizes should be given, but every animal passed by the judges should be entitled to be registered in the Herd Book on the same footing as if it had taken a prize at the principal show.

A numerous meeting of Guernsey farmers, held immediately after these resolutions were passed, expressed their dissatisfaction with the arrangements thus made. It was felt that while they lowered, or destroyed, the value of honours gained at the island shows, they still subjected any Herd Book founded on them to the mischiefs involved in the fact that selection would be founded on mere external appearance. While the importance of this element was fully admitted, it was considered that, in the case of a breed of which the special excellence lies in the quality as well as the quantity of the milk, exclusive regard to external points was not merely a fallacious test, but one likely to result in deterioration. It was further recognised that many of the most careful breeders in Guernsey had for this very reason never exhibited at the annual shows, and that a book which should exclude their herds would be no fair representative of the best island stock. Instances were recalled in which, both by the judges for the Herd Book in question and by those of the Agricultural Society, grave errors have been made in approving good looking animals of inferior milking quality and of indifferent breed, while rejecting others which in all essential respects were superior. It was held that a check would be given to real progress if the judgment of breeders as to their own stock were to be superseded by that of strangers, and if admission to an official Herd Book was made to depend on conformity to the fancies of a few individuals. It was therefore resolved that any such method of selection of breeding stock would be delusive and injurious, and that the best course was to establish a General Herd Book for the Island of Guernsey, based simply on the principles on which those of Great Britain and America are founded, and therefore open to all stock which might be proved to possess the qualification of purity of race. From such a Register it was believed that breeders would be able to select the strains which they deemed best suited to their purpose, and purchasers would not be misled by any alleged superiority beyond what they had the means of verifying. Rules for the formation of such a Herd Book were approved by the meeting, and a Committee was appointed to superintend its preparation. The First Part is now submitted to the public.

The Committee will only add that the experience even of the first year has shown that owners do not enter their stock indiscriminately, but that in nearly all instances they limit themselves to those which they consider to be their best. If the pedigrees given are short it does not indicate that care has not been bestowed on breeding hitherto, but arises from the fact that it has not been customary to keep a written record. So also the notice of honours obtained by the progenitors is frequently omitted, for the double reason that, not being greatly esteemed, no note has been preserved of them, and that as the stock exhibited at the island shows have not been even distinguished by names till within the last two years, it would not be possible to ascertain with certainty whether honours have been awarded or not. But the Committee feel confident that the Herd Book now inaugurated will lead to the general preservation of more accurate records of breeding, and to the fuller recognition, both in the island and elsewhere, of the value of strains of which judicious breeding has already in numerous instances laid the foundation.

The colours of the Guernsey breed include white, red and black, in any mixture and shade, except roan, no instance of which is known to have occurred. Brindle is not uncommon, The nose may be either white or black.

RULES OF THE GENERAL HERD BOOK.

I.—A book shall be opened which shall be entitled "The General Herd Book of the Island of Guernsey." It shall be the property of and under the direction of the Committee appointed under Rule XII, and the entries in it shall be printed and sold as the Committee may determine.

II.—Every person shall be entitled to enter in such Herd Book any Bull, Cow or Heifer, belonging to him, provided the animal to be entered fulfils all the following conditions :—

    a. It must have been born in Guernsey.

    b. The sire and dam must have been born in Guernsey.

    c. It must be at least one year old, and within Guernsey at the time of entry.

III.—The entries shall be made in forms supplied by the Committee, setting forth the above facts and such other particulars as the Committee may direct, and signed by the actual owner of the animal as true to the best of his knowledge and belief.

IV.—Every animal entered shall bear a name not previously entered, unless it be the direct descendant of an animal already entered, in which case it may bear the same name with addition of a progressive number.

V.—The entries shall be inserted in the Herd Book, and be numbered in the order in which, when fully complying with the preceding rules, they are delivered to the Committee or to any person appointed by it to receive them. If delivered at the same time they shall be entered in alphabetical order of the owners' names.

VI.—Calves may be entered under the entry of their dam, whether born before or after her entry, the date of birth, the sex, and the colour being stated, but no name shall be given to them till entered and numbered separately under Rule II.

VII.—Any honours gained at any of the Guernsey Agricultural Society's Shows, or at any Exhibition out of Guernsey which the Committee may decide to be admissible, may be stated in the entry or inserted afterwards.

VIII.—The Committee may at any time require or admit evidence in respect to any animal entered, and may expunge or correct any entry which it deems inaccurate. Any person proved to the satisfaction of the Committee to have knowingly made a false statement in any entry signed by him, shall not be permitted to enter any more animals.

IX.—There shall be paid for each entry, except those of calves under Rule VI, the sum of 1s. Certified copies of any entry may be obtained on payment of 1s. for each copy.

X.—The Committee may appoint the Secretary or other officer to superintend the keeping of the Herd Book, and may pay to him a suitable sum per annum, and make such regulations for keeping the Book as it thinks fit.

XI.—The Committee may appoint a Sub-Committee of Management of the Herd Book, and such Sub-Committee shall have all the powers hereinbefore conferred on the Committee, or such powers as the Committee may confer. Any decision of the Sub-Committee by which any individual is aggrieved may be reviewed by the Committee.

XII.—A President, Vice-President, Secretary and Committee shall be appointed for the present year by the meeting now held, and in future years by an Annual General Meeting to be held in January of each year; but vacancies in these offices occurring in the course of any year shall be filled up by the Committee till the next Annual Meeting.

XIII.—General Meetings may be called at any time by notice in two preceding numbers of the *Gazette*, signed by the President, Vice-President, or any three Members of the Committee. At all Meetings the right to be present or to vote shall be limited to those persons who, during the year then current, or during the immediately preceding year, have entered stock in the Herd Book. Each person shall have one vote, and the President of the Meeting shall only vote in case of equality in the other votes.

XIV.—Any alteration in the foregoing rules shall only be made by resolution of a General Meeting, notice being given in the advertisement calling such Meeting of the alteration proposed.

REGISTER OF COWS AND HEIFERS.

N.B.—The figures within parenthesis refer to the Registry of Animals.

———o———

1.—ROSÉE, red, born April, 1875, property of PETER GORE MOLLET, Villette, St. Martin ; bred by owner.

Dam, Cherry, red and white, belonging to A. Gallienne, Càches. *Grand-dam,* Cherry, pale red, belonging to Jean De Putron, Fosse. *Sire,* Rouge, belonging to Henry Mauger, Pompe.

Produce.—July, 1878 : Heifer (2), red, by bull of Thomas Fallaize ; retained by owner. July, 1879 : Bull-calf, killed fat. December, 1880 : Heifer, pale red, by bull of Jean Mauger.

2.—ROSÉE 2ND, red, born 2nd July, 1878, property of PETER GORE MOLLET, Villette, St. Martin ; bred by owner.

Dam, Rosée (1). *Sire,* Rouge, belonging to Thomas Fallaize.

3.—SOUSIGNE, red and white, born March, 1876, property of PETER GORE MOLLET, Villette, St. Martin ; bred by owner.

Produce.—July, 1878 : Heifer, pale red, by bull of Thomas Fallaize, killed. August, 1879 : Heifer, red, by bull of Nicolas Jehan ; retained by owner. 1st January, 1881 : Bull-calf by bull of Abraham Gallienne ; retained by owner.

4.—TOPSY, pale red, born March, 1880, property of PETER GORE MOLLET, Villette, St. Martin ; bred by owner.

Dam, Pecunde, pale red, belonging to owner. *Sire,* pale red bull, belonging to Jean Mauger Page.

5.—ROSY, pale red, born June, 1874, property of BONAMY BROUARD, Courtil au Préel, St. Andrew ; bred by Jean Brouard, Courtil au Préel.

Dam, Rosy, pale red and white, belonging to Jean Brouard. *Sire,* bull belonging to Thomas Maindonald, Grand Belle, St. Andrew.

6.—TINY, pale red and white, born September, 1879, property of BONAMY BROUARD, Courtil au Préel, St. Andrew ; bred by owner.

> *Dam*, Tiny, pale red and white, belonging to Jean De Garis. *Sire*, Billy, belonging to J. G. Lenfestey.

7.—BUTTE, red and white, born October, 1879, property of BONAMY BROUARD, Courtil au Préel, St. Andrew ; bred by owner.

> *Dam*, Beauty, pale red and white, belonging to owner. *Sire*, bull belonging to Alfred Mansell.

8.—FANNY, pale yellow, born April, 1878, property of BONAMY BROUARD, Courtil au Préel, St. Andrew ; bred by Daniel Roper, Huriaux.

> *Dam*, a pale red cow, belonging to Daniel Roper. *Sire*, St. Andrew 2nd, belonging to Robert Best, St. Andrew.

9.—FLORA, red, born February, 1876, property of BONAMY BROUARD, Courtil au Préel, St. Andrew ; bred by owner.

> *Dam*, Marthe, pale red, belonging to Jean Brouard. *Sire*, a bull belonging to Alfred Mansell.
> *Produce.*—May, 1880 : Heifer, pale red, by bull of James James, Les Vauxbelets.

10.—ROUGETTE, red and white, born March, 1873, property of HENRY L. C. LANGLOIS, Ferme des Martins, St. Martin ; bred by M. Langlois, St. Sampson.

> *Dam*, red and white cow, belonging to M. Langlois.

11.—BRUNETTE, brown, born 14th April, 1878, property of HENRY L. C. LANGLOIS, Ferme des Martins, St. Martin ; bred by — Clarke, St. Andrew.

> *Dam*, pale red cow, belonging to — Clarke, St. Andrew. *Sire*, red bull, belonging to Robert Best, St. Andrew.
> *Produce.*—15th July, 1880 : Heifer, red, by bull of James Martel, St. Andrew.

12.—JUDY, pale red, born 5th April, 1876, property of HENRY L. C. LANGLOIS, Ferme des Martins, St. Martin ; bred by Charles Norman, Les Martins.

> *Dam*, pale red cow, belonging to C. Norman. *Sire*, Billy, red, belonging to Nicholas Jehan, St. Martin.
> *Produce.*—12th February, 1881 : Heifer, red and white, by bull of H. Gallienne, Câche, St. Martin.

13.—PAYSANS, red and white, born January, 1874, property of HENRY L. C. LANGLOIS, Ferme des Martins, St. Martin ; bred by N. P. LE ROY, Castel.

Dam, red cow, belonging to J. Mansell, Paysans, St. Peter-in-the-Wood. *Sire*, bull of J. Mansell, Paysans.

14.—LADY BOVILL, red and white, born 12th February, 1878, property of HENRY L. C. LANGLOIS, Ferme des Martins, St. Martin ; bred by G. Langlois, St. Martin.

Dam, red and white cow, belonging to G. Langlois. *Sire*, pale red bull, belonging to N. Heaume, Forest.

15.—DAISY, pale red, born 13th June, 1878, property of HENRY L. C. LANGLOIS, Ferme des Martins, St. Martin ; bred by G. Langlois, St. Martin.

Dam, Daisy, pale red, belonging to G. Langlois. *Grand-dam*, pale red cow, belonging to — Parsons, St. Martin. *Sire*, red bull of R. Best, St. Andrew.

16.—QUEEN, red and white, born 1st January, 1878, property of HENRY L. C. LANGLOIS, Ferme des Martins, St. Martin ; bred by Mrs. Smith, St. Martin.

Dam, red and white, belonging to Mrs. Smith. *Sire*, red bull of J. Martel, Castel.

17.—NELLY, red, born August, 1874, property of HENRY L. C. LANGLOIS, Ferme des Martins, St. Martin ; bred by Mrs. Smith, St. Andrew.

Dam, red cow, belonging to Mrs. Smith. *Sire*, red bull of — Parsons, St. Martin.

18.—BLUEBELL, red and white, born January, 1879 ; property of HENRY L. C. LANGLOIS, Ferme des Martins, St. Martin ; bred by — Priaulx, Castel.

Dam, red and white cow, belonging to — Ozanne. *Sire*, red bull, belonging to — Ozanne.

19.—VIOLET, yellow and white, born 1869, property of JOHN BOYD KINNEAR, Courtil Rozel, St. Peter-Port ; bred at Castel Hospital.

Dam, cow belonging to Castel Hospital. *Sire*, red and white bull of James Martel, Préel, Castel.

Produce.—8th September, 1871 : Bull, sold. 22nd November, 1872 : Heifer, Violet 2nd (20). 11th September, 1873 : Bull, sold. 26th November, 1874 : Bull, sold. 23rd December, 1875 : Bull, sold. 23rd February, 1877 : Bull, sold. 1st May, 1878 : Heifer, Violet 3rd (21). 9th May, 1879 : Heifer, Violet 4th (22). 3rd July, 1880 : Heifer, Violet 7th (25).

20.—VIOLET 2ND, yellow and white, born 22nd November, 1872, property of JOHN BOYD KINNEAR, Courtil Rozel, St. Peter-Port ; bred by owner.

Dam, Violet (19). *Sire*, bull of James James, Vauxbelets, 1st Prize Guernsey, 1871.

Produce.—24th April, 1874 : Bull, sold. 18th May, 1876 : dead. 28th May, 1877 : Heifer, sold. 10th October, 1878 : Heifer, sold. 4th October, 1879 : Bull, Seigneur (Bulls 1). 12th December, 1880 : Bull, Seneschal (Bulls 3). 23rd February, 1882 : Heifer, Violet 10th, by bull of T. Mahy, Calais.

21.—VIOLET 3RD, cream colour and white, born 1st May, 1878, property of JOHN BOYD KINNEAR, Courtil Rozel, St. Peter-Port ; bred by owner.

Dam, Violet (19). *Sire*, bull of J. De Garis, Foulon.

Produce.—9th May, 1880 : Heifer, Violet 6th (24). 17th July, 1881 : Heifer, Violet 8th, by Bailiff (Bulls 2).

22.—VIOLET 4TH, cream colour and white, born 9th May, 1879, property of JOHN BOYD KINNEAR, Courtil Rozel, St. Peter-Port ; bred by owner.

Dam, Violet (19). *Sire*, bull of J. De Garis, Foulon.

Produce.—14th September, 1881 : Heifer, Violet 9th, by Bailiff (Bulls 2).

23.—VIOLET 5TH, yellow and white, born 12th May, 1879, property of JOHN BOYD KINNEAR, Courtil Rozel, St. Peter-Port ; bred by owner.

Dam, Violet 2nd (20). *Sire*, bull of R. Best, St. Andrew.

Produce.—14th June, 1881 : Bull, dead.

24.—VIOLET 6TH, red and white, born 9th May, 1880, property of JOHN BOYD KINNEAR, Courtil Rozel, St. Peter-Port ; bred by owner.

Dam, Violet 3rd (21). *Sire*, Cato, belonging to James Martel, Préel, 1st Prize Guernsey,

25.—VIOLET 7TH, red and white, born 3rd July, 1880, property of JOHN BOYD KINNEAR, Courtil Rozel, St. Peter-Port ; bred by owner.

Dam, Violet (19). *Sire*, bull of R. Best, St. Andrew.

26.—SNOWDROP, red and white, born 1870, property of JOHN BOYD KINNEAR, Courtil Rozel, St. Peter-Port ; bred by Elias Guerin, St. Peter-Port.

Dam, red and white cow, belonging to E. Guerin.

Produce.—1st April, 1874 : Bull, sold. 18th May, 1875 : Bull, sold. 21st June, 1876 : Heifer, Snowdrop 2nd (27). 4th June, 1877 : Bull, died. 4th October, 1878 : Bull, sold. 1st October, 1879 : Bull, Bailiff (Bulls 2).

27.—SNOWDROP 2ND, red and white, born 21st January, 1876, property of JOHN BOYD KINNEAR, Courtil Rozel, St. Peter-Port ; bred by owner.

Dam, Snowdrop (26). *Sire*, bull of T. Calais, St. Martin.

Produce.—29th July, 1878 : Heifer, Snowdrop 3rd (28). 6th August, 1879 : Heifer, sold. September, 1880 : Bull, sold. 25th November, 1881 : Heifer, Snowdrop 4th, by Seigneur (Bulls 1).

28.—SNOWDROP 3RD, red and white, born 29th July, 1878, property of JOHN BOYD KINNEAR, Courtil Rozel, St. Peter-Port ; bred by owner.

Dam, Snowdrop 2nd (27). *Sire*, bull of D. Naftel, St. Andrew.

Produce.—10th September, 1880 : Bull, sold.

29.—HAREBELL 2ND, fawn, born 24th February, 1877, property of JOHN BOYD KINNEAR, Courtil Rozel, St. Peter-Port ; bred by owner.

Dam, Jeannette, red and white ; bred by owner ; sold to H. Merriam, Cherrybrook Farm, U.S. *Grand-dam*, Harebell ; bred by owner, out of his cow Crummie, by Billy, bull of O. De Putron. *Sire*, bull bred by owner, out of Alderney cow.

Produce.—30th November, 1879 : Heifer, Harebell 4th (31). 11th March, 1881 : Bull, sold.

30.—HAREBELL 3RD, red and white, born 24th April, 1877, property of JOHN BOYD KINNEAR, Courtil Rozel, St. Peter-Port ; bred by owner.

Dam, Harebell ; bred by owner, out of Crummie, by bull of O. De Putron ; sold to R. B. Wood, Godwell, Ivy Bridge, Devon. *Sire*, red and white bull of T. Maindonald, Grand Belle, Prize R.A.S. of England.

Produce.—30th November, 1879 : Heifer, sold. 21st November, 1880 : Bull, sold. 9th November, 1881 : Heifer, Harebell 5th, red and white.

31.—HAREBELL 4TH, cream colour and white, born 11th March, 1881, property of JOHN BOYD KINNEAR, Courtil Rozel, St. Peter-Port ; bred by owner.

Dam, Harebell 2nd (29). *Sire*, bull of H. B. Ogier, Bernauderie, Castel.

32.—VESTA, red and white, born 1st March, 1882, property of CHARLES SMITH & SON, Caledonia Nursery ; bred by T. Maindonald, Grand Belle, St. Andrew.

Dam, Mary, property of T. Maindonald. *Grand-dam*, Signora, property of P. Mollet, Forest. *Sire*, Billy 2nd, 1st Prize Guernsey, 1872; 2nd Prize R.A.S. of England, 1873; 1st Prize Bath and West of England, 1873; out of Brechette, belonging to T. Maindonald. *Grand-sire*, Billy 1st, 1st Prize Bath and West of England, 1869; 1st Prize Guernsey, 1870; out of Fan, belonging to N. Mellish, St. Martin.

Produce.—April 6, 1876: Bull, killed. April 25, 1877: Vesta 2nd (33). May 1st, 1878: Vesta 3rd (34). April 27th, 1879: Bull, killed. April 24th, 1880: Vesta 4th, fawn and white, by Premier, belonging to James Hocart, Vale, 1st Prize Guernsey. May 1st, 1881: Bull, Vulcan 2nd, by Vulcan (Bulls 4).

33.—VESTA 2ND, fawn and white, born 25th April, 1877, property of CHARLES SMITH & SON, Caledonia Nursery; bred by owner.

Dam, Vesta (32). *Sire*, Charlie, belonging to — Mahy, Rouvets, Vale, 1st Prize Guernsey.

Produce.—November 25th, 1879: Bull, killed. November 3rd, 1880: Heifer, died. November 10, 1881: Bull, killed.

34.—VESTA 3RD, fawn and white, born 1st May, 1878, property of CHARLES SMITH & SON, Caledonia Nursery.

Dam, Vesta (32). *Sire* Charlie,, belonging to — Mahy, Rouvets, Vale, 1st Prize Guernsey.

Produce.—30th October, 1880: Bull, killed. November 20, 1881: Vesta 6th, by Vulcan (Bulls 4).

35.—JUNO, fawn and white, born 1st February, 1873, property of CHARLES SMITH & SON, Caledonia Nursery; bred by H. Giffard, Braye du Valle.

Dam, Princess, red and white, Prize in Guernsey. *Grand-dam*, Buttercup, red and white, both belonging to breeder. *Sire*, Prince of Orange, belonging to J. A. Ogier, Duveaux, 3rd prize Guernsey, 1872.

Produce.—17th October, 1875: Bull, killed. 27th December, 1876: Bull, killed. 20th November, 1877: Bull, killed. 21st April, 1879: Vulcan (Bulls 4). 22nd May, 1880: Heifer, died through accident. 6th May, 1881: Juno No. 3, by Cato, belonging to J. Martel, Préel, Castel, 1st Prize Guernsey.

36.—BARONESS, pale red and white, born 1871, property of ANTHONY ROBERT, Mauxmarquis, St. Andrew; bred by R. Le Poidevin, Croix, St. Peter-in-the-Wood.

Produce.—12th August, 1874: Baroness 2nd (37). 26th August, 1880: Heifer, pale red and white.

37.—BARONESS 2ND, pale red and white, born 12th August, 1874, property of ANTHONY ROBERT, Mauxmarquis, St. Andrew ; bred by owner.

> *Dam,* Baroness (36).
> *Produce.*—18th January, 1877: Baroness 3rd (38). 2nd November, 1879 : Baroness 4th, pale red and white.

38.—BARONESS 3RD, pale red and white, born 18th January, 1877, property of ANTHONY ROBERT, Mauxmarquis, St. Andrew ; bred by owner.

> *Dam,* Baroness 2nd (37).
> *Produce.*—23rd September, 1880 : Heifer.

39.—LADY JANE, red and white, born 1874, property of ANTHONY ROBERT, Mauxmarquis, St. Andrew.

40.—DUCHESS, pale red and white, born 5th September, 1875, property of ANTHONY ROBERT, Mauxmarquis, St. Andrew ; bred by owner.

> *Produce.*—7th September, 1877: Bull, sold to R. Best, St. Andrew. 13th September, 1878: Heifer, sold to J. Bisson, Pleinheaume. 1st December, 1879 : Heifer, sold to Miss Jehan, Torteval. 10th February, 1881 : Bull.

41.—CATHERINE 2ND, pale red, born October, 1864, property of WILLIAM DE LA MARE, Mourants, St. Andrew ; bred by owner.

> *Dam,* Catherine, red and white, belonging to W. De La Mare. *Grand-dam,* cow belonging to Pierre Duquemin. *Sire,* pale red bull, belonging to James Le Page.
> *Produce.*—9th July, 1878 : Catherine 3rd (42). 9th November, 1880 : Heifer, pale red, by Champion, belonging to Abraham Gallienne; retained by owner.

42.—CATHERINE 3RD, red and white, born 9th July, 1878, property of WILLIAM DE LA MARE, Mourants, St. Andrew ; bred by owner.

> *Dam,* Catherine 2nd (41). *Sire,* bull, red and white, of R. Best, St. Andrew.
> *Produce.*—1st January, 1881 : Bull, killed.

43.—GUERNSEY ROSE, red and white, born 30th October, 1877, property of J. A. NAFTEL, Lohiers, St. Saviour ; bred by owner.

> *Dam,* red and white cow, belonging to owner. *Sire,* pale red and white bull of A. Blondel, Neuve Maison.

44.—FLOWER, red, born 25th April, 1877, property of J. A. NAFTEL, Lohiers, St. Saviour ; bred by owner.

Dam, red cow of owner. *Grand-dam*, red cow of owner. *Sire*, bull of Alfred Mansell, Villiaze, 3rd Prize Guernsey.

45.—ANNIE,   , born 15th May, 1879, property of J. A. NAFTEL, Lohiers, St. Saviour ; bred by owner.

Dam and *Grand-dam*, both belonging to owner. *Sire*, pale red and white bull of P. De La Mare, Prevost.

46.—MERRY MAID, red, born April, 1878, property of J. A. NAFTEL, Lohiers, St. Saviour ; bred by owner.

Dam, red and white cow, belonging to W. F. Allez, Forest. *Grand-dam*, cow belonging to George Corbin, Douït, St. Peter-in-the-Wood. *Sire*, bull of Francis Jehan, L'Alleur, Torteval.

47.—LUCY, red and white, born 3rd July, 1879, property of J. A. NAFTEL, Lohiers, St. Saviour ; bred by owner.

Dam, red and white cow, belonging to Thomas Lenfestey, Fontaines. *Grand-dam*, cow belonging to Josiah N. Le Prevôts. *Sire*, bull of Alfred Mansell, St. Andrew, 4th Prize Guernsey.

48.—NELLY, red and white, born April, 1880, property of J. A. NAFTEL, Lohiers, St. Saviour ; bred by owner.

Dam, pale red and white cow, belonging to owner. *Grand-dam*, cow belonging to owner. *Sire*, bull of M. Lainé, Issues.

49.—LADY FLORA, pale red, self, born 30th March, 1880, property of LOUISE NAFTEL, Héchet, Castel ; bred by owner.

Dam and *Grand-dam*, cows belonging to Daniel Ogier, Marais.

50.—ROSETTE, red and white, born May, 1879, property of LOUISE NAFTEL, Héchet, Castel ; bred by owner.

Dam, red and white cow, belonging to P. Martel, Pièces, Forest. *Grand-dam*, cow belonging to J. Brouard, Vinaires.

51.—BIJOU, pale red, born 7th May, 1879, property of LOUISE NAFTEL, Héchet, Castel ; bred by owner.

Dam, very pale cream colour cow, belonging to Jean Corbin, Adams. *Grand-dam*, cow belonging to owner. *Sire*, red and white bull, belonging to James Paint, Fontaines, by bull belonging to same.

52.—PANSY, cream colour, born May, 1876, property of
LOUISE NAFTEL, Héchet, Castel ; bred by J. Mansell,
Paysans.

> *Dam*, cream colour and white cow belonging to breeder.
> *Grand-dam*, cow, belonging to same.
> *Produce.*—29th September, 1880 : Pansy 2nd, pale red, by
> bull of Jean Priaulx, Friquet, Castel (3rd Prize Guernsey) ;
> retained by owner.

53.—GUERNSEY LILY, pale red, born 25th December,
1877, property of LOUISE NAFTEL, Héchet, Castel ;
bred by W. Le Ruez, St. Andrew.

> *Dam*, pale red and white cow, belonging to breeder. *Grand-
> dam*, cow, belonging to Misses De Garis, Raies. *Sire*, bull of
> Alfred Mansell, Villiaze (3rd Prize Guernsey), by bull of said
> owner.

54.—DAIRY MAID, red and white, born 1871, property of
JEAN BROUARD, Choffins, St. Saviour ; bred by Mrs.
Tostevin, Villiaze, St. Andrew.

> *Produce.*—July, 1879 : Flora, Heifer, red and white. No-
> vember, 1880 : Bull, red and white, by bull of C. Le Ray.

55.—FAIRY, red, born August, 1871, property of JEAN
BROUARD, Choffins, St. Saviour ; bred by owner.

> *Dam*, Jessie, red and white, belonging to owner. *Sire*, bull
> of Abraham Robin.
> *Produce.*—March, 1874 : Heifer, Guernsey Lily, pale red,
> by bull of A. Robin, sold for America. February, 1875 : Bull,
> by bull of Jean Brouard, sold to — Sausmarez. February,
> 1876 : Heifer, died. 1st April, 1877 : Bull, red and white, by
> bull of A. Mauger, sold. August, 1878 : Bull, died. June,
> 1879 : Heifer, red and white, by bull of A. Robin. Septem-
> ber, 1880 : Bull, by bull of N. P. Le Roy.

56.—BUTTERFLY, red and white, born 1874, property of
JOHN BROUARD ; bred by Mme. Tostevin, Villiaze, St.
Andrew.

> *Produce.*—February, 1880 : Bull, Jack, red and white, by
> bull of R. Best, St. Andrew. January, 1881 : Bull, red and
> white, by bull of — Le Ray.

57.—ROBINE, fawn, born 14th June, 1879, property of
DENYS CORBET, La Roberge, Forest ; bred by James
Gilmore, Bourg, Forest.

> *Dam*, Robinotte, dark fawn and white, belonging to breeder.
> *Sire*, bull of N. Heaume, Houards, Forest, Prize in Guernsey.

B

58.—CHÂTELAINE, pale red and white, born 25th September, 1879, property of DENYS CORBET, La Roberge, Forest ; bred by James Gilmore, Bourg, Forest.

Dam, cow belonging to breeder. *Sire,* bull of N. Heaume, Forest.

59.—MARIAN, pale red and white, born 1872, property of DENYS CORBET, La Roberge, Forest ; bred by George Corbin, Douït, St. Peter-in-the-Wood.

Dam, cow belonging to breeder.
Produce.—31st March, 1881 : Heifer.

60.—POLLY, fawn and white, born 12th June, 1876, property of DENYS CORBET, La Roberge, Forest ; bred by Josué Badier, Fontaine, St. Sampson.

Dam, fawn cow, belonging to Rev. J. Watson, St. Martin. *Grand-dam,* cow of Josué Badier. *Sire,* bull of R. Best, St. Andrew.
Produce.—October, 1878 : Heifer, pale red and white, by bull of N. Heaume, Houards, 3rd Prize Guernsey, sold to — De Garis, Grés, St. Peter-in-the-Wood. September, 1879 : Bull, fawn and white, by bull of T. Maindonald, Grand Belle, Prize Guernsey. October, 1880 : Bull, fawn and white, by bull of H. Ogier, Bernauderie, 1st Prize Guernsey.

61.—BIJOU, pale red and white, born November, 1879, property of PIERRE FALLA, Pulias, St. Sampson ; bred by Daniel Henry, Vardes, St. Sampson.

62.—STELLA, red and white, born 1873, property of THOMAS LE POIDEVIN, Barras, Vale ; bred by owner.

Sire, bull of Thomas Maindonald, Grand Belle, St. Andrew.
Produce.—April 1881 : Heifer, by bull of Thomas Mahy, Rouvets, Vale.

63.—LEILA, nearly red, born 1876, property of THOMAS LE POIDEVIN, Barras, Vale ; bred by owner.

Sire, bull of Thomas Mahy, Rouvets, Vale.

64.—LEILA 2ND, red and white, born 1879, property of THOMAS LE POIDEVIN, Barras, Vale ; bred by owner.

Dam, Leila (63).

65.—REINE D'OR, red and white, born 1875, property of THOMAS LE POIDEVIN, Barras, Vale ; bred by owner.

Sire, bull of Charles Le Page, Naftiaux, St. Andrew, 1st Prize Guernsey.

66.—REINE D'OR 2ND, pale red and white, born 24th May, 1880, property of THOMAS LE POIDEVIN, Barras, Vale ; bred by owner.

Dam, Reine d'Or (65).

67.—LADY DIANA, red and white, born 1873, property of NICOLAS FALLA, Courtil Jacques, Vale ; bred by owner.

Dam, red and white cow, belonging to owner.

68.—LADY DIANA 2ND, red and white, born 1875, property of NICOLAS FALLA, Courtil Jacques, Vale ; bred by owner.

Dam, Lady Diana (67).

69.—LADY DIANA 3RD, red and white, born December, 1879, property of NICOLAS FALLA, Courtil Jacques, Vale.

70.—NANNY, red, born 9th March, 1876, property of RICHARD MAHY, La Passée, St. Sampson ; bred by owner.

Dam, red cow, belonging to owner. *Sire*, bull, red and white, of John Ozanne, Genâts, Castel, Commended Guernsey.

71.—PASSÉE, pale red and white, born March, 1874, property of RICHARD MAHY, Passée, St. Sampson ; bred by owner.

Dam, pale red and white cow of owner. *Sire*, pale bull of owner.

72.—PASSÉE 2ND, pale red and white, born April, 1877, property of RICHARD MAHY, Passée, St. Sampson ; bred by owner.

Dam, Passée (71). *Sire*, red and white bull of owner, Commended Guernsey.

73.—LA GRANDE, pale red and white, born 1882, property of DANIEL HENRY, Les Vardes, St. Sampson ; bred by Colonel H. Giffard, Braye, St. Sampson.

Dam, pale red and white cow, belonging to Colonel H. Giffard.

74.—LA GRANDE 2ND, pale red and white, born 1877, property of DANIEL HENRY, Les Vardes, St. Sampson ; bred by owner.

Dam, La Grande (73). *Sire*, pale red bull of Nicolas Robin, Grand Clos, Castel.

Produce.—July, 1880 : Bull, red and white, by bull of Nicolas Mahy, Rouvets, Vale.

75.—LA GRANDE 3RD, pale red and white, born February, 1880, property of DANIEL HENRY, Les Vardes, St. Sampson ; bred by owner.

> *Sire*, red and white bull of Thomas Hocart, Petite Hougue, Vale, Commended Guernsey.

76.—LOUISE, pale red and white, born March, 1879, property of THOMAS H. LAINÉ, Vaugrat, St. Sampson; bred by owner.

> *Dam*, pale red and white cow belonging to owner. *Sire*, bull of Thomas Mahy, Rouvets, Vale.

77.—LA NORMANDIE, red, born 1874, property of THOMAS H. LAINÉ, Vaugrat, St. Sampson ; bred by owner.

> *Dam*, brindle cow of owner.

78.—LA NORMANDIE 2ND, red, born March, 1878, property of THOMAS H. LAINÉ, Vaugrat, St. Sampson ; bred by owner.

> *Dam*, La Normandie (77).
> *Produce.*—1st April, 1881 : Heifer, red, by bull of Thomas Mahy, Rouvets, Vale.

79.—LA NORMANDIE 3RD, red, born April, 1879, property of THOMAS LAINÉ, Vaugrat, St. Sampson.

> *Dam*, La Normandie (77). *Sire*, red and white bull of Jean Robin, Longs Camps, St. Sampson.

80.—NONNEY, red, born 1879, property of THOMAS J. HENRY, Houmet, Vale ; bred by Jean Martin, La Ruette, St. Martin.

81.—LADY JEANNETTE, red and white, born 1879, property of THOMAS J. HENRY, Houmet, Vale ; bred by Jean Robin, Longs Camps, St. Sampson.

82.—PICOTTE, pale red, born 1877, property of THOMAS LE POIDEVIN, Martins, St. Sampson ; bred by owner.

> *Dam*, pale red cow of owner. *Sire*, bull of Jacques Le Patourel, Ramée, St. Peter-Port.

83.—PICOTTE 2ND, pale red, born 17th January, 1880, property of THOMAS LE POIDEVIN, Martins, St. Sampson ; bred by owner.

> *Dam*, pale red cow of owner. *Sire*, bull of Thomas Mahy, Rouvets, Vale.

84.—FLORETTE, red and white, born 22nd June, 1879, property of THOMAS LE POIDEVIN, Martins, St. Sampson.

> *Dam*, red and white cow of owner. *Sire*, bull of Thomas Mahy, Rouvets, Vale.

85.—ROUGE, red, born May, 1878, property of THOMAS LE POIDEVIN, Martins, St. Sampson ; bred by Jacques Le Patourel, Ramée, St. Peter-Port.

86.—JUILLETTE, red and white, born 19th July, 1879, property of THOMAS LE POIDEVIN, Martins, St. Sampson ; bred by owner.

> *Dam*, red and white cow of owner. *Sire*, bull of Thomas Mahy, Rouvets, Vale.

87.—CAPELLES, red and white, born 1874, property of THOMAS MAHY, Capelles, St. Sampson ; bred by Peter Le Poidevin, Pulias, Vale.

88.—ROSEA, pale red and white, born November, 1879, property of THOMAS MAHY, Capelles, St. Sampson ; bred by Thomas Ogier, Tertre, St. Sampson.

89.—BLUE GEM, born March, 1880, red and white, property of THOMAS MAHY, Capelles, St. Sampson ; bred by Josué Sebire, Emrais, Castel.

90.—MARIE LOUISE, red, born 1876, property of THOMAS MAHY, Capelles, St. Sampson ; bred by — Le Patourel, St. Martin.

91.—AGNES, red and white, born 8th August, 1879, property of GEORGE ALLEZ, Buttes, St. Saviour ; bred by owner.

> *Dam*, Marie, red and white cow of owner. *Grand-dam*, cow of John and George Allez, Le Bordage, St. Saviour. *Sire*, red and white bull of — Hocart.

92.—CECILIA, light red and white, born 1st June, 1879, property of GEORGE ALLEZ, Buttes, St. Saviour ; bred by owner.

> *Dam*, Curly, light red and white cow of owner. *Grand-dam*, cow of John and George Allez, Bordage, St. Saviour. *Sire*, bull of — Hocart, Hougue du Pommier, Castel, 1st Prize Guernsey.

93.—HENRIETTE, fawn, born 1877, property of GEORGE ALLEZ, Buttes, St. Saviour ; bred by John and George Allez, Le Bordage, St. Saviour.

Dam, light red cow of breeders. *Grand-dam*, red cow of breeders.

Produce.—13th June, 1879 : Bull, Henry, by bull of H. De Garis, sold to Mr. Maxwell. 3rd September, 1880 : Heifer, Henriette 2nd, by St. Andrew 3rd, belonging to R. Best, St. Andrew, 1st Prize Guernsey.

94.—CURLY, light red and white, property of GEORGE ALLEZ, Buttes, St. Saviour ; bred by John and George Allez, Le Bordage, St. Saviour.

Dam, light red and white cow of breeders. *Grand-dam*, light red and white cow of Nicolas Allez.

Produce.—June, 1878 : Heifer, Geraldine, light red and white, sold to Charles Le Page. 1st June, 1879 : Heifer, Cecilia, light red and white, by bull of — Hocart, 1st Prize Guernsey. 15th June, 1880 : Heifer, Curly 2nd, light red and white, by St. Andrew 3rd, belonging to R. Best, St. Andrew, 1st Prize Guernsey.

95.—CURLY 2ND, light red and white, born 15th June, 1880, property of GEORGE ALLEZ, Buttes, St. Saviour ; bred by owner.

Dam, Curly (94). *Sire*, St. Andrew 3rd, bull of R. Best, St. Andrew, 1st Prize Guernsey.

96.—SUSIE, light red, property of GEORGE ALLEZ, Buttes, St. Saviour ; bred by John and George Allez, Le Bordage, St. Saviour, 5th Prize Guernsey, 1880.

Dam, red cow of breeders. *Grand-dam*, cow of James Alexandre, Hamel, St. Saviour.

Produce.—1877 : Heifer, Henriette (93). 25th November, 1878 : Bull, died. 21st May, 1880 : Bull, Henry, by Billy, bull of T. G. Lenfestey, 3rd Prize Guernsey.

97.—FLORA, pale red, born April, 1879, property of GEORGE TORODE, Bourg, Forest ; bred by owner.

Dam, Dairy Maid, 2nd Prize Guernsey. *Sire*, Billy, property of N. Heaume, Forest, 4th Prize Guernsey.

98.—STAR, pale red, born August, 1879, property of GEORGE TORODE, Bourg, Forest ; bred by owner.

Dam, Victor, red cow of owner, 1st Prize Guernsey. *Sire*, Billy, red and white, bull of N. Heaume, 3rd Prize Guernsey.

99.—PERI, red and white, born May, 1879, property of GEORGE TORODE, Bourg, Forest ; bred by owner.

Dam, Dairy Maid, pale red cow of owner. *Sire*, Billy, pale red bull of N. Heaume, 4th Prize Guernsey.

100.—ZILLA, red, born January, 1879, property of GEORGE TORODE, Bourg, Forest ; bred by N. Priaulx, Grantez, Castel.

101.—JUNO DU BOURG, red, born March, 1877, property of GEORGE TORODE, Bourg, Forest ; bred by owner.

Sire, Billy, pale bull of S. Best, 1st Prize Guernsey.

102.—VENUS, pale red, born May, 1877, property of GEORGE TORODE, Bourg, Forest ; bred by owner.

Dam, Dairy Maid, pale red cow of owner, 2nd Prize Guernsey. *Sire*, Billy, pale red bull of N. Heaume, 3rd Prize Guernsey.

103.—VASHTI, pale red, born May, 1878, property of GEORGE TORODE, Bourg, Forest ; bred by owner.

Dam, Dairy Maid, pale red cow of owner, 2nd Prize Guernsey. *Sire*, Billy, pale red bull of N. Heaume, 3rd Prize Guernsey.

104.—VICTOR, red, born March, 1868, property of GEORGE TORODE, Bourg, Forest ; bred by owner.

Dam, Star, pale red cow of owner. *Sire*, pale red bull of N. Heaume, 3rd Prize Guernsey.

105.—LOTTIE, pale red, property of JOHN LE HURAY, Croûtes, St. Peter-in-the-Wood ; bred by P. LANGLOIS, St. Peter-in-the-Wood.

106.—MARQUISE, pale red and white, born 1875, property of DANIEL MARQUIS, Landes, Castel ; bred by Thomas Le Poidevin, Barras, Vale.

107.—STE. PIERRAISE, white and red, 1876, property of DANIEL MARQUIS, Landes, Castel ; bred by James Langlois, Hèche, St. Peter-in-the-Wood.

Produce.—March, 1881 : Heifer, by bull of Henry Ogier, Bernauderie, Castel.

108.—MARIE, pale red, property of JOHN LE HURAY, Croûtes, St. Peter-in-the-Wood ; bred by P. Langlois, St. Peter-in-the-Wood.

109.—DAISY DE LA VILLE BAUDU, red and white, born June, 1874, property of THOMAS H. OGIER, Ville Baudu, Vale ; bred by owner.

Dam, Fanny (115). *Sire*, red and white bull of owner, *Produce.*—29th September, 1879 : Daisy de la Ville Baudu 2nd (110). ———— : Bull, Eclipse, sold to R. Best, Prize Guernsey.

110.—DAISY DE LA VILLE BAUDU 2ND, red and white, born 19th September, 1879, property of THOMAS H. OGIER, Ville Baudu, Vale ; bred by owner.

Dam, Daisy de la Ville Baudu (109). *Sire,* Billy, pale red bull of owner, out of Dairy Maid, pale red cow of owner.

111.—SMUTTY, pale red, born June, 1874, property of THOMAS H. OGIER, Ville Baudu, Vale ; bred by owner.

Dam, Brownie, property of owner. *Grand-dam,* cow of — Sarchet. *Sire,* pale red bull of J. A. Ogier.

112.—DUCHESSE DE LA VILLE BAUDU, red and white, born 13th June, 1879, property of THOMAS H. OGIER, Ville Baudu, Vale ; bred by owner.

Dam, Dairy Maid, red and white cow of owner. *Sire,* red and white bull of J. Collas, Maison de Bas.

113.—DOLLY, pale red, born June, 1878, property of THOMAS H. OGIER, Ville Baudu, Vale ; bred by owner.

Dam, Fanny (115). *Sire,* Eclipse, red and white bull of owner, Prize Guernsey.

114.—BIJOU DE LA VILLE BAUDU, brown, born January, 1879, property of THOMAS H. OGIER, Ville Baudu, Vale ; bred by Mrs. Hubert, Clos-du-Lande, Vale.

Dam, Rose du Clos, pale red cow of breeder. *Grand-dam,* Grande Fleurie, cow of James Falla. *Sire,* red and white bull of J. Collas.

115.—FANNY, pale red and white, born 1871, property of THOMAS H. OGIER, Ville Baudu, Vale ; bred by Mrs. Ogier, Tertre.

Dam, red and white cow of Mrs. Ogier. *Grand-dam,* red cow of Mrs. Ogier. *Sire,* red and white bull of Thomas Ogier.
Produce.—June, 1874 : Daisy de la Ville Baudu (109). June, 1878 : Dolly (113). 29th July, 1880 : Champion (Bulls 7).

116.—NANCY, red and white, born February, 1877, property of THOMAS H. OGIER ; bred by James Hubert, Parc.

117.—NANCY DAWSON, spotted red and white, born March, 1880, property of THOMAS P. BICHARD, Millbrook, Vrangue ; bred by G. Ash, Bouët.

Dam, red and white cow of breeder.

118.—TOPSY DE LA ROCQUE BALAN, pale red, born February, 1875, property of W. M. Collas, Rocque Balan, Vale ; bred by owner.

Dam, pale red and white cow of owner. *Sire*, bull of Thomas H. Ogier, Ville Baudu.

Produce.—29th June, 1881 : Heifer, pale red and white.

119.—POLLY DE LA ROCQUE BALAN, pale red, born 1876, property of W. M. Collas, Rocque Balan, Vale ; bred by owner.

Dam, brindled cow of owner. *Sire*, bull of John A. Ogier, Devaux, St. Sampson.

120.—EDNA, pale red, born August, 1878, property of Thomas Collas, Marais, Vale ; bred by Daniel Gavet, Hautgard, Vale.

Dam, pale red, belonging to Thomas Hocart, Petites Hougues. *Sire*, Premier, belonging to Thomas Hocart, out of cow belonging to Misses Harvey, Rouge Huis.

121.—LA HAUTGARD, pale red, born August, 1875, property of Daniel Gavet, Hautgard, Vale ; bred by owner.

Sire, pale red bull, out of cow belonging to Misses Harvey, Rouge Huis.

Produce.—March, 1881 : Heifer, red and white.

122.—LALANDAISE, pale red and white, born October, 1879, property of Pierre Mollet, Les Landes, Clos du Valle ; bred by W. M. Collas, Rocque Balan, Vale.

123.—ROSE, pale red, born 1876, property of Mary Hubert, Clos des Landes, Vale ; bred by owner.

Dam, pale red and white cow belonging to owner. *Sire*, bull of Thomas H. Ogier, Ville Baudu, Vale.

Produce.—June, 1880 : Rose 2nd (124). July, 1881 : Heifer, pale red.

124.—ROSE 2ND, pale red and white, born June, 1880, property of Mary Hubert, Clos des Landes, Vale ; bred by owner.

Dam, Rose (123). *Sire*, bull belonging to Job Henry, L'Hermitage, Vale.

125.—LILY DU CLOS DES LANDES, pale red, born 1876, property of Mary Hubert, Clos des Landes, Vale ; bred by owner.

Dam, pale red and white cow belonging to owner. *Sire*, bull belonging to Thomas Ogier, Ville Baudu.

Produce.—Lily du Clos des Landes 2nd (126).

c

126.—LILY DU CLOS DES LANDES 2ND, pale red and white, born December, 1879, property of MARY HUBERT, Clos des Landes, Vale ; bred by owner.

Dam, Lily du Clos des Landes (125). *Sire*, Bismarck, bull belonging to Thomas Ogier, Ville Baudu, Prize Guernsey.

127.—ROSITTA DU ROSEWOOD, fawn and white, born 1877, property of NICOLAS GUILBERT, Rosewood, Castel ; bred by owner.

Dam, cow belonging to Daniel Le Gallais, St. Saviour. *Produce.*—Rositta 2nd (128). January, 1881 : Rositta 3rd, fawn and white, by bull of Mrs. Hocart, Hougue du Pommier.

128.—ROSITTA DU ROSEWOOD 2ND, red and white, born July, 1879, property of NICOLAS GUILBERT, Rosewood, Castel ; bred by owner.

Dam, Rositta (127). *Sire*, bull belonging to Nicolas Robin, Grand Clos, Castel.

129.—PRINCESS, red and white, born 8th August, 1880, property of NICOLAS GUILBERT, Rosewood, Castel ; bred by owner.

Dam, pale red and white cow belonging to Jacques Le Patourel, Ramée, St. Peter-Port. *Sire*, bull belonging to Mrs. Hocart, Hougue du Pommier, Castel.

130.—CHEMINANTE 3RD, dark red, born 1877, property of MATTHEW TOSTEVIN, Delisles, Castel ; bred by owner.

Dam, Cheminante 2nd, dark red, belonging to owner. *Grand-dam*, Cheminante, dark red, belonging to owner. *Sire*, red and white bull belonging to James Martel, Préel, Castel, 2nd Prize Guernsey, by bull belonging to same owner. *Produce.*—Cheminante 4th (131). August, 1881 : Heifer, red.

131.—CHEMINANTE 4TH, pale red and white, born 1879, property of MATTHEW TOSTEVIN, Delisles, Castel ; bred by owner.

Dam, Cheminante 3rd (130). *Sire*, red and white bull belonging to James Martel, Préel, Castel, by bull belonging to same owner.

132.—ADAMS, pale red, born 1876, property of MATTHEW TOSTEVIN, Delisles, Castel ; bred by Maurice Adams, L'Erée Hotel.

133.—PRIMY, red and white, born 1877, property of NICOLAS PRIAULX, Fauxquets de Bas, Castel, bred by owner.

Dam, pale red cow belonging to owner. *Produce.*—Primy 2nd (134). January, 1881 : Heifer, red and white, by bull of R. Dumaresq, Moulin de Haut, Castel.

134.—PRIMY 2ND, pale red, born March, 1879, property of NICOLAS PRIAULX, Fauxquets de Bas, Castel ; bred by owner.

Dam, Primy (133). *Sire,* bull belonging to Thomas Maindonald, Grand Belle, St. Andrew.

135.—JANE 2ND, red and white, born 1874, Prize Guernsey, property of JEAN OGIER, La Croix, Castel ; bred by owner.

Dam, Jane, pale red cow belonging to owner. *Sire,* bull belonging to James Martel, Préel, Castel.

Produce.—Jane 3rd (136).

136.—JANE 3RD, pale red, born 1879, property of JEAN OGIER, La Croix, Castel ; bred by owner.

Dam, Jane 2nd (135). *Sire,* Jimmy, pale red bull belonging to Henri Ogier, Bernauderie, Prize Guernsey.

Produce.—8th September, 1881 : Heifer, Jane 4th, red and white, by bull belonging to Jacques Le Page, Neuve Maison, Castel.

137.—ELIZABETH, red and white, born 1877, property of JEAN MARTEL, Haut Pavé, Castel ; bred by owner.

Dam, red and white cow belonging to Pierre Martel, Mare, Castel. *Sire,* red and white bull belonging to owner.

138.—NANCY DU HAUT PAVÉ, red, born 1873, property of JEAN MARTEL, Haut Pavé, Castel ; bred by J. Fleure, Maison d'Aval, Vale.

Produce.—March, 1881 : Heifer, red, by bull belonging to Pierre Martel, Mare, Castel.

139.—LA REINE, pale red, born 1873, property of JEAN MARTEL, Haut Pavé, Castel ; bred by E. Mahy, Haute Lande, Vale.

Produce.—January, 1881 : Bull, Royal Quand-même, pale red, by bull belonging to Pierre Martel, Mare, Castel.

140.—CÉSARÉE, red and white, born 1869, property of JEAN MARTEL, Haut Pavé, Castel ; bred by owner.

Produce.—1877 : Césarée 2nd (141). May, 1881 : Heifer, red and white, by bull belonging to Pierre Martel, Mare, Castel.

141.—CÉSARÉE 2ND, red and white, born 1877, property of JEAN MARTEL, Haut Pavé, Castel ; bred by owner.

Dam, Césarée (140). *Sire,* red and white bull belonging to owner.

142.—EMILY, brindled, born 1874, property of DANIEL H. DOREY, Plaisance, Castel ; bred by owner.

Dam, brindled cow belonging to owner. *Sire*, bull belonging to James Martel, Préel, Castel.
Produce.—Emily 2nd (143).

143.—EMILY 2ND, pale red, born 1876, property of DANIEL H. DOREY, Plaisance, Castel ; bred by owner.

Dam, Emily (142). *Sire*, bull belonging to James Martel, Préel, Castel.
Produce.—August, 1880 : Heifer, pale red, by bull belonging to W. Carrington, Grands Moulins, Castel.

144.—JESSY, red and white, born 12th July, 1875, property of THOMAS LE PELLEY, Les Videclins, Castel ; bred by owner.

Dam, pale red cow belonging to owner. *Sire*, bull belonging to James Martel, Préel.
Produce.—Jessy 2nd (145).

145.—JESSY 2ND, red and white, born June, 1880, property of THOMAS LE PELLEY, Les Videclins, Castel ; bred by owner.

Dam, Jessy (144). *Sire*, red and white bull belonging to owner.

146.—BEAUTY, red, born 1878, property of PIERRE LE CHEMINANT, Douït, Castel ; bred by owner.

Dam, red cow belonging to owner. *Grand-dam*, red cow belonging to the same. *Sire*, red bull belonging to owner, by red bull belonging to the same.

147.—LILY, red, born 1879, property of PIERRE LE CHEMINANT, Douït, Castel ; bred by owner.

Dam, red cow belonging to owner. *Grand-dam*, red cow belonging to the same, Prize Guernsey. *Sire*, red bull belonging to owner.

148.—LADY DES BEAUCAMPS, pale red and white, born 1873 ; property of THOMAS N. LE PELLEY, Beaucamps de Haut, Castel ; bred by owner.

Dam, pale red and white cow belonging to owner. *Grand-dam*, pale red and white cow belonging to the same. *Sire*, bull belonging to James Martel, Préel.
Produce.—25th June, 1881 : Lady des Beaucamps 2nd (149).

149.—LADY DES BEAUCAMPS 2ND, red and white, born 25th June, 1881 ; property of THOMAS N. LE PELLEY, Beaucamps de Haut, Castel ; bred by owner.

Dam, Lady des Beaucamps (148). *Sire*, red and white bull belonging to James Martel, Préel.

150.—VICTORIA, pale red and white, born June, 1873, property of THOMAS N. LE PELLEY, Beaucamps de Haut, Castel ; bred by Mrs. J. Le Pelley, Beaucamps de Haut.

Dam, red and white cow belonging to Mrs. J. Le Pelley, Beaucamps de Haut.

Produce.—7th January, 1881 : Victoria 2nd (151).

151.—VICTORIA 2ND, red and white, born 29th December, 1881, property of THOMAS N. LE PELLEY, Beaucamps de Haut, Castel ; bred by owner.

Dam, Victoria (150). *Sire*, Cato, red and white, belonging to James Martel, Préel, 1st Prize Guernsey.

152.—LA PERLE, red and white, born 1873, property of THOMAS N. LE PELLEY, Beaucamps de Haut, Castel ; bred by James Martel, Préel, Castel.

Sire, bull belonging to James Martel, Préel.

Produce.—La Perle 2nd (153). 25th April, 1881 : Heifer, red and white, by bull belonging to James Martel, Préel, 1st Prize Guernsey.

153.—LA PERLE 2ND, red and white, born 20th April, 1879, belonging to THOMAS N. LE PELLEY, Beaucamps de Haut, Castel ; bred by owner.

Dam, La Perle (152).

154.—TULIP, red and white, born 1875, property of E. A. BEST, Villiaze, St. Andrew.

Dam, black and white cow belonging to owner. *Sire*, red and white bull belonging to R. Best, Briqueterie, St. Andrew.

155.—CHERRY, red and white, born 1879, property of E. A. BEST, Villiaze, St. Andrew ; bred by owner.

Dam, red and white cow belonging to Mrs. Parsons, St. Martin.

156.—BUTTERCUP, black, born 1878, property of E. A. BEST, Villiaze, St. Andrew ; bred by owner.

Dam, black and white cow belonging to T. Fallaize, St. Martin ; bred by him.

Produce.—1880 : Heifer, black, by bull belonging to — Heaume, Les Houards.

157.—ROSE DE LA VILLIAZE, red, born 1879, property of E. A. BEST, Villiaze, St. Andrew ; bred by owner.

Dam, brindled cow belonging to R. Poidevin, Capelles. *Sire*, bull belonging to Thomas Maby, Capelles, Prize Guernsey.

158.—FAUVETTE, pale red, born 1877, property of PETER MARTEL, Hougue Fouque, St. Saviour ; bred by F. Falla, Camps, Castel.

Dam, pale red cow belonging to — Falla, Camps, Castel.

Produce.—1881 : Heifer, pale red, by bull belonging to — Dumaresq.

159.—LILY DE LA HOUGUE FOUQUE, pale red, born 1879, property of PETER MARTEL, Hougue Fouque, St. Saviour ; bred by owner.

Dam, pale red cow belonging to — Gallienne, La Croûte, Torteval.

160.—DAISY DE LA HOUGUE FOUQUE, pale red and white, born 1879, property of PETER MARTEL, Hougue Fouque, St. Saviour ; bred by owner.

Dam, pale red and white cow belonging to — Le Poidevin, Parto, Castel. *Sire*, Billy, pale red and white, belonging to N. Priaulx, Fauxquets.

Produce.—February, 1882 : Heifer, pale red and white.

161.—FAVOURITE, pale red and white, born 1877, property of JEAN A. N. LAINÉ, Lohiers, St. Saviour ; bred by J. De G. Lainé, Lohiers.

Dam, Black Polly, belonging to J. De G. Lainé, Lohiers, St. Saviour. *Grand-dam*, cow belonging to H. De Garis, Les Belles. *Sire*, bull belonging to — Tostevin, St. Saviour.

162.—POLLY DES LOHIERS 2ND, black, born 1876, property of JEAN A. N. LAINÉ, Lohiers, St. Saviour ; bred by J. De J. Lainé.

Dam, Black Polly, belonging to J. De G. Lainé, Lohiers. *Sire*, bull belonging to Le M. Lainé, Les Issues, St. Saviour.

Produce.—1880 : Heifer, pale red and white.

163.—LILY DES BELLES, red and white, born 1879, property of HENRI DE GARIS, Les Belles, St. Saviour ; bred by owner.

Dam, Beauty, belonging to owner. *Grand-dam*, belonging to same. *Sire*, bull belonging to Alfred Mansell, Villiaze.

164.—FLORA DES BELLES, red, born 1875, property of HENRI DE GARIS, Les Belles, St. Saviour ; bred by — Robilliard, St. Peter-in-the-Wood.

Dam, red cow belonging to — Robilliard, St. Peter-in-the-Wood.

Produce.—January, 1882 : Heifer, red.

165.—JOLIE, red and white, born 1880, property of PIERRE DE MOUILPIED, Fosse, St. Martin ; bred by Jesse Dunn, Croûte Havilland.

Dam, red and white cow belonging to Jean De Putron. *Sire*, Billy, belonging to T. Mahy, Calais, St. Martin.

166.—DOLLY DE LA FOSSE, fawn, born April, 1880, property of JOHN LOCOCK, La Fosse, St. Martin ; bred by owner.

Dam, Cherry, belonging to owner. *Sire*, red bull belonging to R. Best, Briqueterie, St. Andrew.

167.—NELLY DE LA FOSSE, red, born February, 1880, property of JOHN LOCOCK, La Fosse, St. Martin ; bred by owner.

Dam, fawn and white cow belonging to owner. *Grand-dam*, fawn and white cow belonging to Miss Mansell, Forest. *Sire*, red bull belonging to R. Best.

168.—PRETTY MAID, red, born March, 1880, property of JOHN LOCOCK, La Fosse, St. Martin ; bred by C. Norman, Villette, St. Martin.

Dam, Pretty Maid, belonging to C. Norman, Villette, St. Martin. *Grand-dam*, Daisy, belonging to same. *Sire*, red bull belonging to R. Best, St. Andrew, by bull belonging to Major Feilden, Herm.

169.—DAISY DE LA FOSSE, fawn and white, born 1878, property of JOHN LOCOCK, La Fosse, St. Martin ; bred by owner.

Dam, fawn cow belonging to owner. *Grand-dam*, cow belonging to Miss Mansell. *Sire*, Cloth of Gold, belonging to Rev. Joshua R. Watson, La Fosse, St. Martin.

170.—LUCY DE L'ÉGLISE, red and white, born 1874, property of Mrs. P. LE PATOUREL, Sous l'Eglise, Castel ; bred by J. Le Page, Hubits, St. Martin.

Dam, red and white cow belonging to — Le Page, Hubits.

171.—ROSA, red and white, born 1879, property of Mrs. P. LE PATOUREL, Sous l'Eglise, Castel ; bred by owner.

Dam, red and white cow belonging to — Hansford. *Sire*, bull belonging to J. Mahy, Rouvets, 2nd Prize Guernsey.

Produce.—1881 : Heifer, red and white, by Cato, belonging to James Martel, Préel, 1st Prize Guernsey.

172.—LADY MARY, red and white, born 1879, property of PIERRE GIRARD, Tertre, Castel ; bred by owner.

Dam, Nancy, belonging to owner. *Grand-dam,* red cow belonging to — Corbin, St. Peter-in-the-Wood. *Sire,* red and white bull belonging to — Le Page, Neuve Maison.

173.—FILLPAIL, black and white, born 1879, belonging to Amos Chick, Albecq, Castel ; bred by owner.

Dam, black cow belonging to owner. *Sire,* red and white bull belonging to James Martel, Préel, Castel.

174.—DAIRY MAID D'ALBECQ 2ND, red and white, born 1876, belonging to Amos Chick, Albecq, Castel ; bred by owner.

Dam, Dairy Maid, belonging to owner. *Sire,* red and white bull belonging to James Martel, Préel, Castel.

Produce.—1881 : Heifer, red and white, by bull of James Martel, Préel.

175.—TINKER 2ND, black and white, born 1878, property of Amos Chick, Albecq, Castel ; bred by owner.

Dam, Tinker, belonging to owner. *Sire,* red and white bull belonging to James Martel, Préel, Castel.

176.—FLORA DES VALLES, pale red and white, born 1879, 4th Prize Guernsey, 1881, property of Richard A. Dumaresq, Valles, Castel ; bred by Mrs. Varrant, Portes des Granges.

Dam, pale red and white cow belonging to Mrs. Varrant. *Sire,* Loyal, pale red and white, belonging to John Davey, St. Peter-Port, 4th Prize Guernsey.

177.—HENRIE, red and white, born 1877, property of John Jehan, Maison d'Aval, St. Sampson ; bred by owner.

Dam, red cow belonging to owner.

Produce.—June 1881 : Heifer, red and white.

178.—APPOLINE, pale red, born 1879, property of Susanne Martin, Grands Moulins, Castel ; bred by N. De Garis, Grands Moulins.

Dam, pale red cow belonging to J. Lenfestey, Ste. Appoline.

179.—FILLPAIL DES GRANDS MOULINS, red and white, born 1870, property of Susanne Martin, Grands Moulins, Castel ; bred by N. De Garis, Grands Moulins.

Dam, pale red cow belonging to — Langlois, St. Peter-in-the-Wood.

Produce.—27th December, 1881 : Heifer, red.

180.—ELISE, black and white, born 1877, property of
SUSANNE MARTIN, Grands Moulins, Castel ; bred by N.
De Garis, Grands Moulins.

> *Dam*, pale red cow belonging to T. Corbin, Varendes.
> *Produce.*—1st December, 1881 : Heifer, pale red.

181.—PASSÉE 3RD, red and white, born 12th March, 1881,
property of RICHARD MAHY, La Passée, St. Sampson ;
bred by owner.

> *Dam*, Passée 2nd (72). *Sire*, Charley, red and white,
> Rouvets, Vale.

182.—JANNENETTE, yellow, born 28th March, 1881, property of RICHARD MAHY, La Passée, St. Sampson ;
bred by owner.

> *Sire*, Bismarck, belonging to Thomas Ogier, Ville Baudu.

183.—POLLY DES PAYSANS, mixed colour, born May,
1879, property of PIERRE TOSTEVIN, Les Paysans, St.
Peter-in-the-Wood.

> *Dam*, black cow belonging to H. Quertier, St. Andrew.
> *Sire*, bull belonging to R. Best, St. Andrew.
> *Produce.*—30th July, 1881 : Heifer.

184.—LILY DES PAYSANS, red, born 9th April, 1878,
property of PIERRE TOSTEVIN, Les Paysans, St. Peter-
in-the-Wood ; bred by owner.

185.—POLLY DU LONG FRIE 3RD, red and white, born
1875, property of T. TOSTEVIN, Long Frie, St. Peter-in-
the-Wood ; bred by owner.

> *Dam*, Polly du Long Frie 2nd, pale red. *Grand-dam*,
> Polly du Long Frie, both belonging to owner. *Sire*, Billy,
> pale red, belonging to owner.

186.—POLLY DU LONG FRIE 4TH, red, born January,
1880, property of T. TOSTEVIN, Long Frie, St. Peter-in-
the-Wood ; bred by owner.

> *Dam*, Polly du Long Frie 2nd. *Grand-dam*, Polly du Long
> Frie, both belonging to owner. *Sire*, Billy, pale red, by Billy,
> both belonging to A. Blondel.

187.—DAISY DU LONG FRIE 3RD, pale red, born 1876,
property of T. TOSTEVIN, Long Frie, St. Peter-in-the-
Wood ; bred by owner.

> *Dam*, Daisy du Long Frie 2nd. *Grand-dam*, Daisy du
> Long Frie, both belonging to owner. *Sire*, Billy, pale red,
> belonging to owner.
> *Produce.*—Daisy du Long Frie 4th (188).

D

188.—DAISY DU LONG FRIE 4TH, pale red, born January, 1881, property of T. TOSTEVIN, Long Frie, St. Peter-in-the-Wood ; bred by owner.

Dam, Daisy du Long Frie 3rd (187).   *Sire*, Billy, pale red, by Billy, both belonging to A. Blondel.

189.—FLOWER DU LONG FRIE, red and white, born 1880, property of T. TOSTEVIN, Long Frie, St. Peter-in-the-Wood ; bred by — Cameron, Jardinet, Castel.

Dam, Fleurie, red and white, belonging to breeder.   *Sire*, Goldfinch, pale red, belonging to Jean Priaulx, Castel.

190.—FLORA DU LONG FRIE, red and white, born January, 1880, property of N. P. TOSTEVIN, Long Frie, St. Peter-in-the-Wood ; bred by James Le Moigne, Pleinmont, Torteval.

Dam, red and white cow belonging to breeder.   *Sire*, Billy, red and white, belonging to W. Brehaut.

191.—CATHERINE DES MOURANTS 4TH, pale red, born 19th November, 1880, property of WILLIAM DE LA MARE, Mourants, St. Andrew ; bred by owner.

Dam, Catherine 3rd, pale red.   *Grand-dam*, Catherine 2nd, red and white, both belonging to owner.   *Sire*, Champion, red and white, belonging to Abraham Gallienne.

192.—JANE DU GREHOGNET, red and white, born 3rd July, 1880, property of JOHN OGIER, Grehognet, Castel ; bred by owner.

Dam, red and white cow belonging to owner.   *Sire*, bull belonging to R. Dumaresq, Castel.

193.—LILY DU GREHOGNET, pale red, born 10th May, 1880, property of JOHN OGIER, Grehognet, Castel ; bred by owner.

194.—TILLY, pale red, born March, 1878, property of THOMAS LENFESTEY, Houguette, St. Peter-in-the-Wood ; bred by owner.

Dam, pale red cow of owner.

Produce.—8th April, 1880 : Bull, pale red, by Sambo, pale red, belonging to owner.

195.—BEAUTÉ, red, born May, 1877, property of THOMAS LENFESTEY, Houguette, St. Peter-in-the-Wood ; bred by owner.

Dam, pale red cow of owner.

196.—ROSE DE LA HOUGUETTE, pale red, born January, 1880, property of Thomas Lenfestey, Houguette, St. Peter-in-the-Wood; bred by owner.

Dam, pale red cow of owner.

197.—JANE DE LA HOUGUETTE, pale red, born April, 1880, property of Thomas Lenfestey, Houguette, St. Peter-in-the-Wood; bred by owner.

Dam, pale red cow of owner.

198.—POLLY DE LA PLAISANCE, red, born May, 1880, property of Nicolas Tostevin, Plaisance, St. Peter-in-the-Wood; bred by owner.

199.—FILASIER, pale red, born January, 1880, property of Nicolas Tostevin, Plaisance, St. Peter-in-the-Wood; bred by owner.

ERRATA.

In Register of Cows,

No. 48, after Nellie, insert des Lohiers.
No. 61, after Bijou, insert du Pulias.
No. 115, after Fanny, insert de la Ville Baudu.
No. 116, after Nancy, insert de la Ville Baudu.
No. 127, after Rositta, delete du Rosewood.
No. 128, after Rositta 2nd, delete du Rosewood.

REGISTER OF BULLS.

———o———

1.—SEIGNEUR, red and white, born 4th October, 1879, property of JOHN BOYD KINNEAR, Courtil Rozel, St. Peter-Port ; bred by owner.

    Dam, Violet 2nd (Cows 20). *Sire,* bull belonging to T. Mahy, Calais, St. Martin.

2.—BAILLIF, red and white, born 1st October, 1879, property of JOHN BOYD KINNEAR, Courtil Rozel, St. Peter-Port ; bred by owner.

    Dam, Snowdrop (Cows 26). *Sire,* St. Andrew 3rd, belonging to R. Best, St. Andrew.

3.—SENESCHAL, red and white, born 12th December, 1881, property of JOHN BOYD KINNEAR, Courtil Rozel, St. Peter-Port ; bred by owner.

    Dam, Violet 2 (Cows 20). *Sire,* bull of — Robin, Grand Clos, St. Saviour.

4.—VULCAN, fawn and white, born 21st April, 1879, property of CHARLES SMITH & SON, Caledonia Nursery, St. Peter-Port ; bred by owners.

    Dam, Juno (Cows 35). *Sire,* Roger, fawn, belonging to — Ozanne, Houmet, Castel, 4th Prize Guernsey, bred by P. Mahy, Les Landes, Vale, by Rival, belonging to T. Ozanne, Blanc Bois, Castel, 3rd Prize Guernsey, out of Brechette, property of T. Maindonald, Grand Belle, St. Andrew.

5.—TAMMY, pale red and white, born April, 1880, property of RICHARD MAHY, Passée, St. Sampson ; bred by owner.

    Dam, Passée (Cows 71). *Sire,* Charlie, red and white, belonging to Thomas Mahy, Rouvets, Vale.

6.—VAUGRAT, red and brown, born March, 1880, property of THOMAS H. LAINÉ, Vaugrat, St. Sampson ; bred by owner.

    Dam, La Normandie (Cows 77), brindled, property of Thomas H. Lainé. *Sire,* bull of Thomas Mahy, Rouvets.

7.—CHAMPION, red and white, born 29th July, 1880, property of THOMAS H. OGIER, Ville Baudu, Vale ; bred by owner.

> *Dam*, Fanny de la Ville Baudu (Cows 115). *Sire*, Charlie, red and white, belonging to Thomas Mahy, Rouvets, Vale, Prize Guernsey.

8.—PRINCE, red and white, born February, 1880, property of THOMAS H. OGIER, Ville Baudu, Vale ; bred by owner.

> *Dam*, Dairy Maid 2nd, pale red and white. *Grand-dam*, Dairy Maid, both property of Thomas H. Ogier. *Sire*, Premier, red and white, belonging to T. Hocart, Petite Hougue.

9.—CHIEF BARON, lemon and white, born June, 1889, property of THOMAS LE PAGE, Villiocq, Castel ; bred by owner.

> *Dam*, Comtesse, property of T. Le Tissier. *Grand-dam*, Princesse, property of John Martel. *Sire*, Cato, belonging to James Martel, Préel, Castel, 1st Prize Guernsey 1880 and 1881, by St. Andrew, belonging to R. Best, St. Andrew, 1st Prize Guernsey 1876 and 1877.

10.—STRANGER, light red, born 1881, property of HENRI DE GARIS, Les Belles, St. Saviour ; bred by J. Burpit, St. Martin.

> *Dam*, red and white cow belonging to Rev. Joshua R. Watson, St. Martin. *Sire*, white and red bull belonging to same.

11.—BILLY, pale red, born 27th July, 1881, property of JOHN OGIER, Grehognet, Castel ; bred by owner.

> *Sire*, bull belonging to — Palmer, Friquet, Castel.

INDEX OF NAMES OF ANIMALS.

N.B.—The numbers refer to the Register of Cows and Heifers, except where the Register of Bulls is specified.

————o————

A.

B.

C.

D.

E.

F.

G.

H.

J.

L.

M.

Marian	59	Marquise	106
Marie	108	Merry Maid	46
Marie Louise	90		

N.

Nancy	70	Nelly	17
Nancy Dawson	117	Nelly de la Fosse	167
Nancy du Haut Pavé	138	Nelly des Lohiers	48
Nancy de la Ville Baudu	116	Nonney	80

P.

Pansy	52	Polly du Long Frie 3rd	185
Passée	71	Polly du Long Frie 4th	186
Passée 2nd	72	Polly des Paysans	183
Passée 3rd	181	Polly de la Plaisance	198
Paysans	13	Polly de la Rocque Balan	119
Peri	99	Pretty Maid	168
Picotte	82	Primy	133
Picotte 2nd	83	Primy 2nd	134
Polly	60	Prince ....(Bulls)	8
Polly des Lohiers	162	Princess	129

Q.

Queen	16

R.

Reine d'Or	65	Rosée	1
Reine d'Or 2nd	66	Rosée 2nd	2
Robine	57	Rosette	50
Rosa	171	Rositta	127
Rose	123	Rositta 2nd	128
Rose 2nd	124	Rosy	5
Rose de la Houguette	196	Rouge	85
Rose de la Villiaze	157	Rougette	10
Rosea	88		

S.

Ste. Pierraise	107	Snowdrop 3rd	28
Seigneur ....(Bulls)	1	Sousigne	3
Seneschal ....(Bulls)	3	Star	98
Smutty	111	Stella	62
Snowdrop	26	Stranger ....(Bulls)	10
Snowdrop 2nd	27	Susie	96

T.

Tammy ....(Bulls)	5	Topsy	4
Tilly	194	Topsy de la Rocque Balan	118
Tinker	175	Tulip	154
Tiny	6		

V.

Z.

INDEX OF OWNERS.

———o———

H.

Henry, Daniel, Les Vardes, St. Sampson, 73, 74, 75.
Henry, Thomas J., Houmet, Vale, 80, 81.
Hubert, Mary, Clos des Landes, Vale, 123, 124, 125, 126.

J.

Jehan, John, Maison d'Aval, St. Sampson, 177.

K.

Kinnear, John Boyd, Courtil Rozel, St. Peter-Port, 19, 20, 21, 22, 23,
24, 25, 26, 27, 28, 29, 30, 31 ; Bulls, 1, 2, 3.

L.

Lainé, Jean A. N., Lohiers, St. Saviour, 161, 162.
Lainé, Thomas H., Vaugrat, St. Sampson, 76, 77, 78, 79 ; Bulls, 6.
Langlois, Henry L. C., Ferme des Martins, St. Martin, 10, 11, 12, 13,
14, 15, 16, 17, 18.
Le Cheminant, Pierre, Douït, Castel, 146, 147.
Le Huray, John, Croûtes, St. Peter-in-the-Wood, 105, 108.
Lenfestey, Thomas, Houguette, St. Peter-in-the-Wood, 194, 195,
196, 197.
Le Page, Thomas, Villiocq, Castel, Bulls, 9.
Le Patourel, Mrs. P., Sous l'Eglise, Castel, 170, 171.
Le Pelley, Thomas, Les Videclins, Castel, 144, 145.
Le Pelley, Thomas N., Beaucamps de Haut, Castel, 148, 149, 150,
151, 152, 153.
Le Poidevin, Thomas, Barras, Vale, 62, 63, 64, 65, 66.
Le Poidevin, Thomas, Martins, St. Sampson, 82, 83, 84, 85, 86.
Locock, John, La Fosse, St. Martin, 166, 167, 168, 169.

M.

Mahy, Richard, La Passée, St. Sampson, 70, 71, 72, 181, 182.
Mahy, Thomas, Capelles, St. Sampson, 87, 88, 89, 90.
Marquis, Daniel, Landes, Castel, 106, 107.
Martel, Jean, Haut Pavé, Castel, 137, 138, 139, 140, 141.
Martel, Peter, Hougue Fouque, St. Saviour, 158, 159, 160.
Martin, Susanne, Grands Moulins, Castel, 178, 179, 180.
Mollet, Peter Gore, Villette, St. Martin, 1, 2, 3, 4.
Mollet, Pierre, Les Landes, Clos du Valle, Vale, 122.

N.

Naftel, J. A., Lohiers, St. Saviour, 43, 44, 45, 46, 47, 48.
Naftel, Louise, Héchet, Castel, 49, 50, 51, 52, 53.

O.

Ogier, Jean, La Croix, Castel, 135, 136.
Ogier, John, Grehognet, Castel, 192, 193 ; Bulls, 11.
Ogier, Thomas-H., Ville Baudu, Vale, 109, 110, 111, 112, 113, 114,
115, 116 ; Bulls, 7, 8.

P.

Priaulx, Nicolas, Fauxquets de Bas, Castel, 133, 134.

R.

Robert, Anthony, Mauxmarquis, St. Andrew, 36, 37, 38, 39, 40.

S.

Smith, Charles, and Son, Caledonia Nursery, St. Peter-Port, 32, 33, 34, 35 ; Bulls, 4.

T.

Torode, George, Bourg, Forest, 97, 98, 99, 100, 101, 102, 103, 104.
Tostevin, Matthew, Delisles, Castel, 130, 131, 132.
Tostevin, Nicolas, Plaisance, St. Peter-in-the-Wood, 198, 199.
Tostevin, Pierre, Les Paysans, St. Peter-in-the-Wood, 183, 184.
Tostevin, T., Long Frie, St. Peter-in-the-Wood, 185, 186, 187, 188, 189, 190.

THOMAS M. BICHARD, Printer, Bordage Street, Guernsey.

www.ingramcontent.com/pod-product-compliance
Lightning Source LLC
Chambersburg PA
CBHW060006230526
45472CB00008B/1963